OFF-SHORE OIL
RIG WORKERS

BY
Gail Stewart

EDITED BY
Anita Larsen

PUBLISHED BY
CRESTWOOD HOUSE
Mankato, MN, U.S.A.

CIP

LIBRARY OF CONGRESS CATALOGING IN PUBLICATION DATA

Stewart, Gail.
  Off-shore oil rig workers.

  (At risk)
  Includes index.
  SUMMARY: Describes the important and often hazardous job of off-shore oil rig workers.
  1. Off-shore oil industry—Employees—Juvenile literature. [1. Off-shore oil industry—Employees. 2. Occupations.] I. Larsen, Anita. II. Title. III. Series.
  HD8039.O34S74   1988   622'.3382—dc19   88-12006
  ISBN 0-89686-397-2

International Standard Book Number:
0-89686-397-2

Library of Congress Catalog Card Number:
88-12006

PHOTO CREDITS

Cover: Tom Stack & Associates: Steve Ogden
Tom Stack & Associates: (Steve Ogden) 4, 22, 29, 34; (Jim McNee) 7
Journalism Services: (Don Allen) 11, 14, 16, 17, 19, 30-31, 37
Third Coast Stock Source: (Regis Lefebure) 13; (Paul H. Henning) 42-43
Wide World Photos, Inc.: 20, 26
DRK Photo: (Stephen J. Krasemann) 23, 25, 39

Copyright © 1988 by Crestwood House, Inc. All rights reserved. No part of this book may be reproduced in any form without written permission from the publisher, except for brief passages included in a review. Printed in the United States of America.

Produced by Carnival Enterprises.

CRESTWOOD HOUSE

Box 3427, Mankato, MN, U.S.A. 56002

TABLE OF CONTENTS

A Man-Made Island................................5
The Importance of Oil............................6
How Oil Was Made.................................8
The Demand Increases.............................9
The Continental Shelf............................9
Where There's a Will............................10
Drill Where?....................................12
Types of Drilling Rigs..........................15
Platforms.......................................17
Flotel..18
Roughnecks and Derrickmen.......................19
Round Trip......................................22
Mud...24
Roustabouts.....................................27
Divers Under Pressure...........................32
Snoopy..35
Getting To Work.................................36
No Place for Dieters!...........................38
Blowout!..38
Other Dangers...................................40
Why Do This Work?...............................41
For More Information............................44
Glossary/Index...............................45-47

Oil rig platforms are man-made islands built miles from shore.

A MAN-MADE ISLAND

More than 200 miles from shore sits a huge silver platform. It is cluttered with tall yellow towers, cranes, and pipes. On one end of the platform is a smokestack with a long, orange flame burning sideways from it, like a candle tipped at an angle. Waves and winds batter the platform, which creaks and groans but stands firm.

Workers move everywhere about this platform. Some are operating gigantic machines on the surface of the platform. Some are inside it, sleeping or eating. Some hover above in helicopters. Some are hundreds of feet beneath it in special diving suits.

Everywhere there is noise—the sharp blasts of machinery, the chopping of the helicopter blades, the moan of the wind, the crash of the ocean waves. And every 30 seconds three long cries of the foghorn warn passing ships, "Stay away...Stay away."

The people who work—and live—on this platform are here for one purpose. They drill for the oil that lies miles below the ocean floor.

Each worker has a job and must do it well. In this little colony everyone depends on everyone else. And with good reason! The man-made island has a "balance" all its own. The skills and abilities of the workers are the only defense against disaster.

On an offshore platform, the possibilities for disaster are many! Fifty-foot waves and hurricane-force winds or the threat of fire and explosion—all are possible no matter how careful the workers are. No wonder the job of oil platform workers is considered an "at-risk" occupation!

THE IMPORTANCE OF OIL

To understand how important offshore oil platforms are, it's important to understand how important oil itself is.

Oil is used to make tens of thousands of different products. Plastics, nylon, toothpaste, paint, fertilizers, rubber, and ink are just a few. The most important use of oil is gasoline. Our world is on the move, and we depend on a good supply of fuel just for our automobiles. In fact, the world uses about 60 million barrels of oil *per day*! Of all the countries of the world, the one that uses the most oil is the United States.

Things would work out fine if we produced as much oil as we needed. We don't. We must rely on other countries—like those in the Middle East—for much of our oil. The Middle East produces a little more than half the oil the world uses. The countries of the Middle East use very little of their own oil, so they have lots to sell.

Whenever these countries decide to drop oil production or raise the price of each barrel of oil they sell, then the U.S. and other countries are caught with a smaller supply of oil.

In 1973 and 1979 that happened. Iran, one of the Middle East oil producers, shut down some of its oil fields for a time. That caused big shortages of oil worldwide. People worried about not having enough gas in their cars. Panicking, they hurried to fill their

Each worker on a platform has his or her own job.

gas tanks. The lines at the gas pumps were long. Many gas stations simply closed down because they couldn't meet the demand. The gasoline situation gradually improved, but the experience taught many countries a lesson. It's better not to be dependent on other countries for necessities like oil. So oil must be found elsewhere.

HOW OIL WAS MADE

Oil in the ground was formed millions of years ago. At that time, shallow oceans covered most of the earth. When small plants and animals died, they sank to the ocean floor. Year after year, they were covered with sand, mud, and more layers of dead plants and animals. As the plants and animals decayed over time, they turned into what we call oil and natural gas. Because they were formed by the remains of living things, oil and natural gas are sometimes called "fossil fuels." The mud and sand gradually became layers of rock surrounding the layers of fossil fuel.

The history of Earth includes many changes in its surface. The shallow oceans were pushed back by mountains and large areas of land. Layers of rock have heaved and shifted. Oil and gas settled in little pockets between these rocks. There it lay, waiting. Finally people figured out what a great source of energy it was.

THE DEMAND INCREASES

More than 200 years ago, people in this country came across oil, usually when they were digging for something else. Some inventors discovered that the oil could be made into lamp oil or used to lubricate machines. In 1859 the first organized oil drilling took place in Pennsylvania. Soon there were many people interested in looking for oil. They found rich supplies of it in states such as Texas, Ohio, Kentucky, and Oklahoma.

The invention of the automobile spurred on those searching for oil. Whoever could quickly supply oil for gasoline could make lots of money. Oil was becoming big business!

THE CONTINENTAL SHELF

At first, all oil drilling took place on land. As new inventions found new uses for oil and natural gas, demand increased. For many years, oil drillers and engineers knew that lots of oil could be found off the coastlines of the United States. The land under the ocean just offshore, up to about 600 feet deep, is called the "continental shelf."

The continental shelf borders every continent. Any country bordered by an ocean has a continental shelf. Knowing that oil is beneath the ocean and doing something about it are two different things. Drilling on land was hard enough, but drilling underwater seemed terribly difficult. It would be expensive, too.

WHERE THERE'S A WILL...

As so often happens, the problem was solved because oil was too valuable a resource to ignore. Oil companies knew that their expensive work would pay off. If plentiful new supplies of oil could be found from the ocean floor, the money spent on exploring would be worth it. Engineers in the oil industry got busy.

At first continental shelf oil was drilled by onshore wells. Holes were angled out toward the sea. Later, drilling equipment was set up on long piers jutting out into the ocean. Some oil was found, but there had to be some way to reach out into deeper waters.

Finally, oil drilling rigs were set up on man-made

Today, there are about 3,000 oil rigs that drill 24 hours a day.

structures far away from the shore. These structures, called "platforms," are anchored in water hundreds of feet deep. They often sit hundreds of miles away from shore! There are close to 3,000 oil rigs drilling today, 24 hours a day. They supply almost half of the oil we use.

DRILL WHERE?

The continental shelf is a vast area. An oil company can't just set up a platform and start drilling anywhere. Platforms and oil rigs cost many millions of dollars. It would be foolish to assemble a drilling site without first making sure oil was beneath it.

Geologists are people who study the surface of the earth. They understand what types of rocks are good indications of an oil field hiding below. Geologists have a number of ways of telling whether an area is promising.

They may use sensitive instruments such as a "magnetometer" and a "gravitometer." These show the kind of rocks below the surface and how hard they are. The instruments show whether or not the rocks are magnetic (oil is rarely found near magnetic rocks).

A "seismic survey" is a common way to find the right kind of rocks for an oil field. A boat tows a special device that sets off small explosions under the water. An instrument called a "seismograph" measures the shock waves from these explosions as they bounce from the sea floor, off rocks, and then upward. The echoes are recorded and charted on a graph.

A geologist assembles all this information into what looks a little like an X-ray of the area. The rock

Drilling is the only way to find out for sure if there is oil underground.

formations may seem just right for an oil deposit. But even with all this technology, there is only one way to find out for sure. Drill!

A jack-up rig is used in shallow water.

TYPES OF DRILLING RIGS

Workers say they are "drilling a wildcat" when they drill an exploratory well. There are several different types of rigs an oil company may use to drill a wildcat.

If the drilling is going to take place in fairly shallow water (less than 300 feet) a "jack-up rig" might be used. A jack-up rig is a huge platform with very long legs. A barge tows this rig into position. Then workers use electric motors to lower the legs until they rest on the ocean floor. Then the platform is jacked up until it is high above the waves. A tall drilling tower, called a "derrick," stands atop the rig.

A "semi-submersible rig" is used in deeper water. Like the jack-up rig, it is towed to the drilling site. Anchors hold it to the right spot while drilling is going on. When drilling is finished, the rig pulls up anchor and can be towed to another drilling spot.

In deep water a "drill ship" may sometimes be used. It, too, has a tall derrick on board for drilling. All the drilling pipes go straight down from the derrick through a special hole in the bottom of the ship. A drill ship puts down its anchor at the drill site, but its engines have to help keep the ship directly above the area. Some new drill ships even use computers to help them during the drilling.

Anchors hold a semi-submersible rig in place while drilling is going on.

Millions of dollars are needed to construct a platform.

All these drilling rigs are expensive. Usually an oil company rents them from a contractor. Some rigs cost $55 million to build. Companies like Shell, Texaco, and Mobil might rent them for a fee of $100,000 per day! And most wildcat drilling doesn't produce even a drop of useable oil. Only about one out of ten wildcats pays off. The other nine are called "dry holes" and are sealed off.

PLATFORMS

Sometimes a wildcat *does* find oil. But an oil company must remain cautious about investing great amounts of money. They may do more tests in the area. How large is the deposit of oil? How deep is it? What is the quality of the oil? Are there minerals mixed with the oil that would make it less pure?

If the tests show the oil field will produce oil for a long time, an oil company will build a "production platform." A production platform is really just a larger, more permanent oil rig. Where an oil drilling rig is mobile—can be towed away once it's done its job—a platform is there to stay. It needs to be strong enough to outlast hurricanes and high waves. It must last for many years, so it can earn back what it cost to build.

FLOTEL

Some words in our language are a combination of two other words. *Brunch,* for example, is a combination of *breakfast* and *lunch*. Workers who spend much of their time on a production platform call them *flotels*. That word is a combination of *floating* and *hotel*!

Some are large. There is living space on some platforms for over 400 workers. They eat in a large dining room and relax in a recreation and lounge area. They can even buy soap and toothpaste in a small shop.

A flotel is far from a hotel, however. Workers are

not there for a vacation. They are there to work. The types of work they do vary widely.

ROUGHNECKS AND DERRICKMEN

The business end of any platform is the derrick. That's the tall tower that supports the weight of the drill. The drill itself is connected to sections of pipe.

Transporting the derrick to the oil rig platform can be tricky.

Each pipe is about ten yards long. These are screwed together to form a long pole. That is how the workers can reach so far under the ocean to drill. This "pole," along with the drill bit at the end, is called the "drill string." An engine turns the whole thing around and around, and the drill bites into the ocean floor.

There are many little things that can make this a slow, tough job. First, drilling under the ocean floor means drilling *way* down. Most of the time, the drills don't find oil until they've gone two miles or more below the sea bottom. So when the string has gone down as far as it can, new ten-foot sections of pipe must be added to make it longer. This is a real team effort!

The head of the drilling crew, called the "tool pusher," shouts orders to the workers, called "roughnecks." These men must be big and strong. Being a roughneck is back-breaking work. The roughnecks use special equipment to unscrew the drill string from the engine before they add new pieces of pipe.

Another worker, called a "derrickman," sits high up on the derrick on a platform called a "monkey-board." He operates lifting gear that will get the next section of pipe ready for the roughnecks to screw onto the string. The pipe is very heavy. It is also oily and slippery. The roughnecks must be careful to avoid mashing their hands or fingers.

Before they add new pieces of pipe, "roughnecks" must unscrew the drill string.

High up on the derrick, the "derrickman" steadies himself on a small platform.

ROUND TRIP

The hardest job the drillers do is changing the drill bit. Drill bits are expensive—about $50,000 apiece—but they are good for only a few hundred yards of drilling. They become worn and dull and don't cut well into the rock below. When this happens, the bit must be replaced.

Each piece of pipe on the drill string has to be unscrewed and set aside as the drill string is pulled back up. Workers call this job a "round trip."

Imagine pulling up 2.5 miles of heavy pipe, ten

Huge wrenches are needed to unscrew each section of pipe.

feet at a time! The derrickman's hoist uses ropes to pull up the pipes from the sea. Huge hydraulic wrenches powered by water are used to unscrew each section of pipe. The roughnecks work in pairs, grabbing the greasy pipes and handing them back to other workers, who stack them neatly on end against the sides of the derrick.

As the workers get closer to the bottom of the drill string, the pipe steams as it comes up. The bottom of the sea may be icy cold, but deep in the earth the rock is boiling hot. The deeper into the ocean floor it travels, the hotter the drill string gets. A drill string at 2.5 miles reaches a temperature of over 200° F. Even after its journey back through the chilly water, it is still hot to the touch.

At last the drill bit itself is retrieved from the hole. It is shaped like a fist, with sharp cogs where the fingers would be. In very hard rock, bits with diamond teeth are used. A diamond is a very hard mineral, and it can cut cleaner and faster than anything made of steel.

The old, used bit looks bent and twisted. Teeth are missing from the cogs. It has taken a real beating against the hard rock below. In comparison, the new bit, painted shiny silver, almost sparkles.

When the bit is replaced, one of the roughnecks may spit twice in the hole. "For luck," he says, laughing. The string is once again lowered into the hole, and the drilling continues.

MUD

Another responsibility of the roughnecks is taking care of the mud pumps. The "mud" isn't mud at all, but a thick, light-brown fluid that is pumped down through the drill string. The pump forces the mud down the pipes and around the bit. It then flows back up the surrounding hole.

The mud does some important jobs. First, it

A thick fluid keeps the bit slippery and cool.

lubricates, or greases, the bit. This helps it slide more easily through the rocks. Second, the mud helps cool down the drill bit. The bit gets hot from the friction caused by grinding hard rock. And, finally, the mud that flows back up to the surface can tell a lot about what is happening below. An expert might be able to see chips of rock or traces of oil or gas in the mud. If the drill string is getting close to a pocket of oil, the mud usually gives the workers advance notice.

ROUSTABOUTS

Another worker on the oil platform is called a "roustabout." Unlike roughnecks, they don't need to be big and muscular. Roustabouts need to be strong and quick. They are the laborers on the platform — sort of all-purpose handymen or handywomen. They must be the type of person who is called a "self-starter." That means that they must be good at figuring out for themselves what needs to be done, then doing it. A group of roustabouts who stood around waiting for someone to give them orders would be useless on an oil platform.

Roustabouts may assist the roughnecks by stacking heavy bags of the dried mud. They may be needed to repair or splice cables. Sometimes they hose down sections of pipe during a round trip, cleaning grease from the threads of each pipe.

Roustabouts are the "all-purpose handymen" on an oil rig platform.

One of the jobs roustabouts never seem to be finished with is painting. It may seem odd that a brand new billion-dollar platform would need to be repainted constantly, but it does. Every day, blasting winds and saltwater pound the platform. This creates rust that can destroy the structure if it isn't properly cared for. A fresh coat of paint, besides making the platform look better, protects the metal against rust and corrosion.

Roustabouts aren't the only painters on a platform. "Scaffolders" have one of the most dangerous jobs there. Theirs is also one of the most spectacular. They work on the girders of the platform below the main deck. Scaffolders are constantly sandblasting away the old paint, priming the metal, and giving it a twice-over paint job.

A scaffolder can't be afraid of heights. He might be dangling from a little cable seat with nothing between him and the water but ten stories of empty space! He puts up temporary platforms, called scaffolds, that look very much like a metal spiderweb. Then he climbs out and begins his lonely, dangerous work.

Derricks are tall—it's a long way down!

Being a roughneck can be back-breaking work.

DIVERS UNDER PRESSURE

Much of the platform is underwater. Maintenance of the platform is a full-time job, and the underwater parts are no different. Sometimes a fishing boat will damage a section of pipe that carries oil to shore. Bolts on the structure can come loose. Barnacles and weeds can get stuck to the legs of the platform. Taking care of these problems is a job for trained divers.

Diving is extremely hazardous work, and the men who do it are highly paid. One of the biggest threats to divers has always been the use of compressed air. Divers rely on tanks of compressed air when they must be underwater for any length of time. The tanks are filled mostly with a gas called nitrogen.

Nitrogen under pressure can present hazards. If a diver is underwater too long, the nitrogen makes him lightheaded or slow to react. He may get blurred vision or grow dizzy. As he rises to the surface, his body may have trouble getting rid of the pressurized gas in his system. When this happens, bubbles of nitrogen form in his bloodstream and cut off the flow of blood to his joints. This is called the "bends."

Getting the bends is an agonizing experience. A diver gets violent, stabbing pains in his back, arms, leg, and stomach. The pain can cause the diver to lose consciousness. It can even kill him.

When the oil companies began building offshore platforms, they knew they would need divers who could remain underwater for a long time.

They spent time and money figuring out new ways to keep divers underwater for eight to ten hours at a stretch. One big change in the last few years has been the new design in divers' living quarters. Instead of bunking with the rest of the workers, divers live in the same pressure as they work in underwater!

In their specially-designed rooms on the lower deck, there is no regular air. Instead, a mixture of oxygen and helium circulates. When the divers are ready to begin work, they step from their special room into a diving bell that is also pressurized. This takes them down to the ocean floor.

For three weeks at a time they never breathe ordinary air. When their three-week shift is over, they spend several days "decompressing." During decompression, they gradually get used to regular air again. Then the divers have four weeks off.

Another danger divers no longer have to fear is icy water. Death from the cold, called "hypothermia," had long been a threat to commercial divers. Under an oil platform, the divers are working at great depths in extremely cold water. Without a new system of keeping warm, a diver would be dead in minutes.

Inventors created a new type of diving suit. Tubes

are sewn into the suit like a series of tiny spiderwebs. Very hot water—around 130° F—is pumped through these tubes. The water comes from the hose that connects the diver to his diving bell. The constant circulation of hot water keeps the diver comfortable as he works.

SNOOPY

Another vehicle besides his diving bell helps a diver. A little plastic ball, called an R.O.V. (remote controlled vehicle), follows each diver underwater. They are operated by supervisors on the platform. The vehicles have television cameras to monitor each diver. Bright worklights on the R.O.V.s guide the divers in the totally black world underwater.

The divers refer to their R.O.V.s as "Snoopy." One young diver who works on a platform in the North Sea explains: "It's cold and dark and lonely down there. Snoopy's like a pet. You take it with you wherever you go. It's company, it's got lights, you can swim for it. Even though you know it can't help, it's still a comfort. It means there is someone there keeping an eye on you. It makes you feel safe."

During cold weather, water lines must be thawed.

GETTING TO WORK

The length of a work shift on an oil platform is usually two weeks. Workers put in 12-hour shifts each day. There are no holidays or vacations for people on the platform. Work goes on 24 hours a day, rain or shine. At the end of the two weeks, the workers have two weeks off ashore. And they have earned those two weeks!

Workers arrive from shore either by helicopter or by supply boat. If they're given a choice, most would rather not take a boat. The huge waves can make anyone with a weak stomach seasick. Besides, unloading people or supplies from a boat is difficult. Large metal baskets, called "birdcages," are used.

A birdcage hangs from a crane on the deck of the platform. The men get inside the birdcage and are lifted onto the platform. Operating a crane takes a sharp eye and amazing reflexes. Trying to set a birdcage on a boat bouncing in the waves can be a nightmare. The workers are always relieved to reach the platform in one piece!

When anyone comes aboard, he or she must sign in. Keeping track of people on a platform is important. Workers have been swept into the sea by a tall wave or a powerful gust of wind, and no one even knew they were gone until too late. On some platforms, the "buddy system" is the rule everyone must follow.

A "birdcage" transports supplies and people from the supply boat to the platform.

After signing in, the workers are assigned a room. Usually there are two or four bunks to a room. While one shift is working, the other is sleeping. Because someone is always sleeping in the room, courtesy is important. Each worker tries to keep his or her part of the room neat and clean. Besides getting a room assignment, each worker picks up a hard hat and boots and a pair of heavy work gloves.

NO PLACE FOR DIETERS!

There isn't much to do on a platform except work and sleep. After a shift is over, a worker may watch a videotaped television show or write a letter home. Most workers say, however, that they're too tired to do anything more strenuous than that.

Sometimes people compare platform life to army life during peacetime. Since workers have so little freedom and life is so confined in the middle of the ocean, mealtime becomes important. In contrast to military food, which has often been called terrible, food on a platform is wonderful.

Workers eat steaks, all kinds of seafood, and roasts. One meal may offer six or seven choices of vegetables and salads. Pies, cakes, and ice cream—all you can eat—are available anytime. Breakfasts are huge—pancakes, French toast, ham, eggs, bacon, and more kinds of cereals than in a grocery store. It's a good thing everyone works so hard, so they can burn off all that food!

BLOWOUT!

Most of the accidents on a platform happen around the little table that turns the drill string. Cuts

Most accidents happen around the table that turns the drill string.

and bruised hands are common, especially to roughnecks and other members of the drill team.

But the problem connected with drilling that oil workers fear the most is a "blowout." A blowout occurs when the flow of oil is so powerful it shoots up in a wild fountain. This usually happens when the drill bit punctures the oil pocket. Such an explosion could kill hundreds of workers. Thousands of gallons of escaping oil also can damage sea life.

One famous blowout happened in 1977 at an oil platform on the North Sea. The platform, called Bravo 14, hit a wild well that blew a fountain of oil 200 feet in the air. Fifty thousand gallons of oil every hour gushed into the water. This posed a terrible danger for the coastlines of Denmark and Norway.

If the well caught fire, the platform and all the workers on it would be lost. Luckily, it did not burn. But by the time the well was brought under control—seven days later—more than 13,000 tons of oil had spilled into the sea.

People like "Red" Adair, a Texan who is world-famous for being able to "cap" (or stop) a wild well, may have the most hazardous job in the world. He and his crew must get close enough to put a four-ton steel "stopper" on the well. To get that close, they often risk being burned alive by a rain of boiling oil and blasts of hot air.

Blowouts are not common. There are devices called blowout preventers that control the pressure of a well being drilled. Even with this equipment, a blowout is a risk that every worker on the platform thinks about sometimes.

OTHER DANGERS

Fire is another risk of platform life. In some ways, the possibility of a fire can be lessened. Smoking is allowed only in certain areas. There are other causes besides a cigarette that can set off a fire. A faulty electric switch or a piece of machinery that overheats can start a fire, too.

Workers are trained to know how to fight fires.

There are fire extinguishers everywhere on the platform. If a fire were burning out of control, workers would use special escape capsules. Each capsule can hold 28 people. Each one has enough fuel to get a safe distance from a burning platform.

Platform supervisors take a keen interest in the dangers of the weather, too. Platforms have been severly damaged in stormy seas, and ocean storms can give little warning. Some oil drilling rigs capsized by wind and waves still float upside down today.

Even though modern platforms are amazingly sturdy, supervisors take no chances. If there is a threat of dangerous weather, such as a hurricane, helicopters fly the workers back to shore.

WHY DO THIS WORK?

With all of these threats, why would anyone work on a platform? Workers are quick to point out that the pay is good—better than similar jobs ashore. For example, construction workers on an oil rig make about one third more than those on land.

But pay isn't the only reason. Most workers enjoy the challenge of their job. The physical work takes its toll, but a warm comradery grows among workers on a platform. They watch out for one another, like sailors on a ship.

Many workers prefer the isolation of platform life to the stresses and strains of shore life.

Some of these people once worked in construction-type jobs on land. Many prefer the isolation to shore life. The stresses and strains of

platform life, they say, are nothing compared to life in a city! Two weeks ashore at a time is enough for them. They are, for the most part, happy where they are.

FOR MORE INFORMATION

For more information about offshore oil rigs and the people who work on them, write to:

Department of the Interior
Minerals Management Service
18th and C St. NW (Room 4210)
Washington, DC 20240

GLOSSARY/INDEX

The Bends 32—A dangerous condition that sometimes affects deep sea divers. Nitrogen bubbles get trapped in the bloodstream, causing intense pain and sometimes death.

Birdcage 36—A large metal basket used to transfer oil rig workers from a boat to the oil drilling platform.

Blowout 39, 40—When a well of oil is so powerful it shoots up in a wild, dangerous explosion.

Continental shelf 10, 12—The land under the ocean from the shore out to a depth of 600 feet. The Continental shelf has been found to be a good spot to drill for oil.

Decompressing 33—The process by which a diver gets used to breathing regular air after breathing compressed air underwater.

Derrick 15, 19, 23—The tall tower from which the drill hangs.

Derrickman 21, 23—A worker who helps the drill crew. The derrickman uses ropes to help pull the heavy drill pipes into position.

Drill bit 22, 24, 27, 39—The part of the drill that actually bites into the rock and makes a hole.

Drill ship 15—A specially designed ship that floats over the drill site and drills for oil.

Drill string 21, 22, 23, 24, 38—The drill bit with all the pipes connected to it.

GLOSSARY/INDEX

Dry hole 17—A well that does not produce oil.

Fossil fuel 8—A substance like oil or natural gas that was formed from the decaying remains of prehistoric plants and animals.

Geologist 12—A scientist who studies the rocks that make up the earth's crust to find a likely spot for oil wells.

Hypothermia 33—Overexposure to cold. Hypothermia can kill a diver not prepared for the bitter cold hundreds of feet below the surface.

Jack-up rig 15—An oil drilling rig used in fairly shallow water. It is towed to the drilling site, and its legs are lengthened until they rest on the bottom.

Monkeyboard 21—The little platform on which the derrickmen sits to work.

Mud 24, 27—A thick brown fluid used to lubricate and cool the drill bit.

Production platform 18—A permanent drilling structure, often hundreds of miles offshore. These platforms are built only when there is proof of a good supply of oil below.

Roughneck 21, 23, 24, 27—A member of the drilling team.

Round trip 22, 27—The lengthy process of hoisting up the drill string and changing the drill bit.

Roustabout 27, 28—A laborer on the oil platform.

GLOSSARY/INDEX

Roustabouts do a wide variety of jobs.

Scaffolder 28—A worker who maintains the outside painted surface of the platform.

Seismic survey 12—One method oil companies use to study the ocean floor. A seismic survey uses explosions measured by sensitive instruments. This tells the geologists the type and formation of rock below the surface.

Semi-submersible rig 15—A drill rig used in deeper water. It is towed to the drill site and anchored to the bottom.

Tool pusher 21—The supervisor of the drilling crew.

Wildcat 15, 17—An exploratory well dug to see if oil lies beneath the surface.